A Gaggle of Geese

ANIMAL GROUPS
ON LAKES & RIVERS

Alex Kuskowski

CONSULTING EDITOR, DIANE CRAIG, M.A./READING SPECIALIST

A Division of ABDO
ABDO
Publishing Company

visit us at www.abdopublishing.com

Published by ABDO Publishing Company, a division of ABDO, P.O. Box 398166, Minneapolis, Minnesota 55439. Printed in the United States of America, North Mankato, Minnesota

062012
092012

 PRINTED ON RECYCLED PAPER

Editor: Liz Salzmann
Content Developer: Nancy Tuminelly
Cover and Interior Design and Production: Anders Hanson, Mighty Media, Inc.
Photo Credits: Shutterstock

Library of Congress Cataloging-in-Publication Data
Kuskowski, Alex.
A gaggle of geese : animal groups on lakes & rivers / Alex Kuskowski.
p. cm. -- (Animal groups)
ISBN 978-1-61783-540-7
1. Lake animals--Behavior--Juvenile literature. 2. Pond animals--Behavior--Juvenile literature. 3. Social behavior in animals--Juvenile literature. I. Title.
QL146.K873 2013
591.763--dc23
 2012009031

SANDCASTLE™ LEVEL: FLUENT

SandCastle™ books are created by a team of professional educators, reading specialists, and content developers around five essential components—phonemic awareness, phonics, vocabulary, text comprehension, and fluency—to assist young readers as they develop reading skills and strategies and increase their general knowledge. All books are written, reviewed, and leveled for guided reading, early reading intervention, and Accelerated Reader® programs for use in shared, guided, and independent reading and writing activities to support a balanced approach to literacy instruction. The SandCastle™ series has four levels that correspond to early literacy development. The levels are provided to help teachers and parents select appropriate books for young readers.

Emerging Readers
(no flags)

Beginning Readers
(1 flag)

Transitional Readers
(2 flags)

Fluent Readers
(3 flags)

Contents

Animals on Lakes & Rivers

Lakes and rivers are bodies of water. They are home to many animals. They are good places to find food.

Some animals live in the water. Other animals live very close to the water.

Why Live in a Group?

Animals often live in groups. Animals in a group can **protect** each other. They can share space, food, and water. They also work together to help raise babies. Many animal groups have fun names!

A Colony of Frogs

Leopard frogs lay their eggs in **breeding** ponds. Sometimes hundreds of frogs end up in one spot. A female frog can lay between 300 and 6,500 eggs!

Frog Names

MALE
male

FEMALE
female

BABY
polliwog, tadpole, froglet

GROUP
colony, army, knot

A Gaggle of Geese

A gaggle of geese includes several sets of parents and their babies. Both parents take care of their young. They feed them and guard them from predators.

Goose Names

MALE
gander

FEMALE
goose

BABY
gosling

GROUP
gaggle, flock

A Cluster of Dragonflies

Dragonflies lay their eggs in or near the water. Nymphs **hatch** from the eggs. The nymphs live in the water for several months. Then they become adult dragonflies.

Dragonfly Names

MALE
king, drake

BABY
nymph

FEMALE
queen

GROUP
cluster, flight

A Raft of Ducks

A raft of ducks travels together to find food. Ducks eat grasses, fish, and **insects** that live in lakes and rivers.

Duck Names

MALE
drake

FEMALE
duck, hen

BABY
duckling

GROUP
raft, sord, flock

An Asylum of Loons

Loons raise their **offspring** on lakes and rivers. The parents swim near their young. Sometimes they carry the baby birds on their backs to keep them safe.

Loon Names

MALE	FEMALE	BABY	GROUP
cock	*hen*	*chick*	*asylum, flock*

A Bale of Turtles

A bale of turtles is more than one turtle. Turtles sometimes leave the water to sit on rocks or logs. They are **cold-blooded**. They bask in the sun to warm up.

Turtle Names

MALE
male

FEMALE
female

BABY
hatchling

GROUP
bale, dule

A Hover of Trout

Trout are fish that live in lakes and rivers. After they are born they **migrate** to other lakes and rivers. But they return every year to where they were born to lay eggs.

Trout Names

MALE	FEMALE	BABY	GROUP
jack	*shedder*	*fry*	*hover, shoal*

More

LAKE & RIVER GROUPS

A cloud of gnats

A shoal of herrings

A bevy of swans

A run of salmon

A skein of mallards

A family of beavers

A bed of clams

Quiz

1. Frogs lay their eggs in ponds. ***True or false?***

2. Geese guard their young from **predators**. ***True or false?***

3. Nymphs do not live in water. ***True or false?***

4. Turtles are **warm-blooded**. ***True or false?***

5. Trout never return to where they were born. ***True or false?***

ANSWERS: 1) TRUE 2) TRUE 3) FALSE 4) FALSE 5) FALSE

Glossary

breeding – related to or used in creating offspring.

cold-blooded – having a body temperature that changes according to the temperature of the surroundings.

hatch – to break out of an egg.

insect – a small creature with two or four wings, six legs, and a body with three sections.

migrate – to move from one area to another, usually at about the same time each year.

offspring – the baby or babies of an animal.

protect – to guard someone or something from harm or danger.

warm-blooded – having a body temperature that is not greatly affected by the temperature of the surroundings.